ANIMAL GENIUSES

CHIMPS

THERESA EMMINIZER

PowerKiDS press

Published in 2026 by The Rosen Publishing Group, Inc.
2544 Clinton Street, Buffalo, NY 14224

First Edition

Editor: Theresa Emminizer
Book Design: Michael Flynn

Photo Credits: Cover, p. 1 Baehaki Hariri/Shutterstock.com; (series background) Viktoriia LiSa/Shutterstock.com; p. 5 Gerdie Hutomo/Shutterstock.com; p. 7 GUDKOV ANDREY/Shutterstock.com; p. 9 Jane Rix/Shutterstock.com; p. 11 ZUMA Press, Inc./Alamy Stock Photo; p. 13 mark higgins/Shutterstock.com; p. 15 ThomasDeco/Shutterstock.com; p. 17 Andy333/Shutterstock.com; p. 19 courtesy of NASA; p. 21 Ondrej Prosicky.

Cataloging-in-Publication Data

Names: Emminizer, Theresa.
Title: Chimps / Theresa Emminizer.
Description: Buffalo, NY : PowerKids Press, 2026. | Series: Animal geniuses | Includes glossary and index.
Identifiers: ISBN 9781499451795 (pbk.) | ISBN 9781499451801 (library bound) | ISBN 9781499451818 (ebook)
Subjects: LCSH: Chimpanzees–Juvenile literature. | Chimpanzees–Behavior–Juvenile literature.| Animal intelligence–Juvenile literature. | Chimpanzees–Psychology–Juvenile literature.
Classification: LCC QL737.P94 E45 2026 | DDC 599.885-dc23

Manufactured in China

CONTENTS

Cute Cousins

Chimpanzees are some of our closest animal relatives, or family members! In fact, chimps share almost nearly all of our genes, the matter that makes us who we are. We share a lot in common with our cute, furry cousins. They're also very clever, or smart!

Chimp Life

Chimps live in the mountains and forests of western and central Africa. They live in large groups called troops. Troops are usually led by older males. There are lots of different chimp families in one troop. Baby chimps stay close to their mothers until they're 7 years old.

Clever Communication

Chimps **communicate** with one another in lots of different ways. They make sounds such as hoots, calls, and grunts to share their feelings. Chimps also clap their hands, make faces, and even hug and kiss! People have even trained some chimps to communicate using **sign language**.

Self-Awareness

Chimps are self-aware, meaning they know who they are! There have been different tests that show chimps are able to **recognize** themselves in a **mirror**. This might not sound special, but it sets them apart from most other kinds of animals. Most animals believe what they see in a mirror is a different being.

Making Tools

Chimps find clever ways to make and use tools to get the things they need. They use sticks to gather bugs for a tasty snack. They use rocks to crack open nuts. Chimps chew up leaves and use them to gather rainwater from inside trees.

Using Medicine

Chimps are so smart they even use **medicine** to treat sickness and cuts! Chimps look for plants they know will help make them well again. They've also been observed, or seen, putting medicine on each other. This is very unusual among animals.

Creative Play

Tools aren't the only things chimps make. They make toys too! Chimps have been seen using fruits like balls to toss and play catch with. They've also been seen playing tug-of-war together. Chimps that live in different places make and use tools differently.

Chimps in Research

Because chimpanzees are smart, highly trainable, and a lot like people, they've often been used for medical testing and other kinds of **research**. On January 31, 1961, a chimp named Ham became the first chimpanzee in space. Many people feel that using chimps for these purposes is unethical, or wrong.

NASA

Chimp Culture

Chimpanzees are some of the smartest animals on Earth. Much like people, they make tools, use medicine, communicate, and play together. Chimps even have their own cultures, or ways of doing things that are passed down from old to young. Chimps can learn and **adapt**, which may be key to their **survival** in times to come.

GLOSSARY

adapt: To change in order to live better in a certain environment.

communicate: To share ideas and feelings through sounds and motions.

medicine: A drug that a doctor gives you to help fight illness.

mirror: A looking glass that reflects an image.

recognize: To see and understand what something is.

research: Careful study to find new knowledge.

sign language: A way to communicate using hands and gestures.

survival: The state of continuing to live.

FOR MORE INFORMATION

BOOKS

Caudron-Robinson, Kit. *Chimpanzees.* Vancouver, B.C.: Engage Books, 2023.

Rose, Rachel. *The 10 Smartest Animals.* Minneapolis, MN: Bearport Publishing Company, 2025.

WEBSITES

Britannica Kids: Chimpanzee
kids.britannica.com/kids/article/chimpanzee/352950
Learn more about amazing chimpanzees.

National Geographic Kids: 10 Chimpanzee Facts
www.natgeokids.com/uk/discover/animals/general-animals/10-chimpanzee-facts/
Find out more fun facts about chimps!

Publisher's note to educators and parents: Our editors have carefully reviewed these websites to ensure that they are suitable for students. Many websites change frequently, however, and we cannot guarantee that a site's future contents will continue to meet our high standards of quality and educational value. Be advised that students should be closely supervised whenever they access the internet.

INDEX